RICE AND THE MAKING OF SOUTH CAROLINA

AN INTRODUCTORY ESSAY

Daniel C. Littlefield
University of Illinois
at Urbana-Champaign

1430 Senate Street (Box 11669)
Columbia, S. C. 29211

Printed on alkaline paper

International Standard Book No.
1-880067-29-3

Produced by the Public Programs Division
Director: Alexia J. Helsley
Editor and designer: Judith M. Andrews
Electronic maps: Tim Belshaw

FOREWORD

For more than four thousand years, humans have been growing and eating rice. The grain probably originated in southeast Asia; the Moors brought it to Spain in the early-8th century; then in 1609, English colonists tried to grow it in Virginia. The story of its introduction into South Carolina is uncertain.

Rice Mill Chimney (ca. 1830–ca. 1860). Laurel Hill Plantation. (South Carolina Department of Archives and History, State Historic Preservation Office.)

At least two traditions persist: one has Dr. Henry Woodward obtaining Madagascar rice from Captain John Turner and introducing it before 1685; the other has the treasurer of the East India Company sending a bag of seed rice to Carolina in 1696. Regardless of the attribution, colonists were growing two varieties in South Carolina by 1696 when the colonial assembly designated rice as one of the commodities that could be used to pay the annual quit rent tax.

By 1718, South Carolinians were exporting 6,773 barrels of rice, each weighing 350 lbs., to England and 2,333 barrels to other colonies.

Originally, Carolinians grew rice on dry land, but early in the eighteenth century, cultivation spread to swampy fresh water areas. By 1748, some planters were damming rain or creek water into reserves to systematically flood their rice fields; in 1758, McKewn Johnstone of Winyah Bay is said to have become the first to use the power of the tides to flood his rice fields with river water; and by 1783, Gideon Dupont of St. James Goose Creek had perfected the tidal culture of rice—a breakthrough that led not only to the more efficient cultivation of the crop but to higher yields with less labor as well.

With the higher yields came a demand for milling—a need Jonathan Lucas met a year or two earlier when he built the state's first water-driven mill on the Santee River. Other tidal-driven mills followed, and in 1817, he and his son constructed the first steam-driven mill in Charleston.

The cultivation of rice required huge capital investments. Planters, who needed a skilled workforce to clear swamps, build and maintain dykes, and to plant, cultivate, harvest, and process the crop, purchased slaves to meet those needs. Thus, the growth of the Carolina rice culture was responsible in many ways for the dramatic increase in the slave population. In Georgetown District throughout the first half of the nineteenth century, for example, slaves made up over 85 percent of the population.

Until the 1850s, rice reigned supreme. But large-scale rice production was limited to the tidal marshes and inland swamps, while cotton became profitable statewide after the invention of the cotton gin. In its heyday, however, rice made a few hundred planters extremely wealthy. It also contributed to cross culturation and the making of Carolina as a rich cultural hybrid. In the lecture that follows, it is this aspect of rice cultivation that Professor Littlefield describes.

Alexia Jones Helsley

PREFACE

This lecture was prepared for South Carolina History Day, 25 October 1994. I based it on my previous work and that of other South Carolina scholars. It is presented with the addition of citations. I wish to thank Mrs. Alexia Jones Helsley of the South Carolina Department of Archives and History for the invitation and for her hospitality and that of other members of the Department during my visit. *Daniel C. Littlefield*

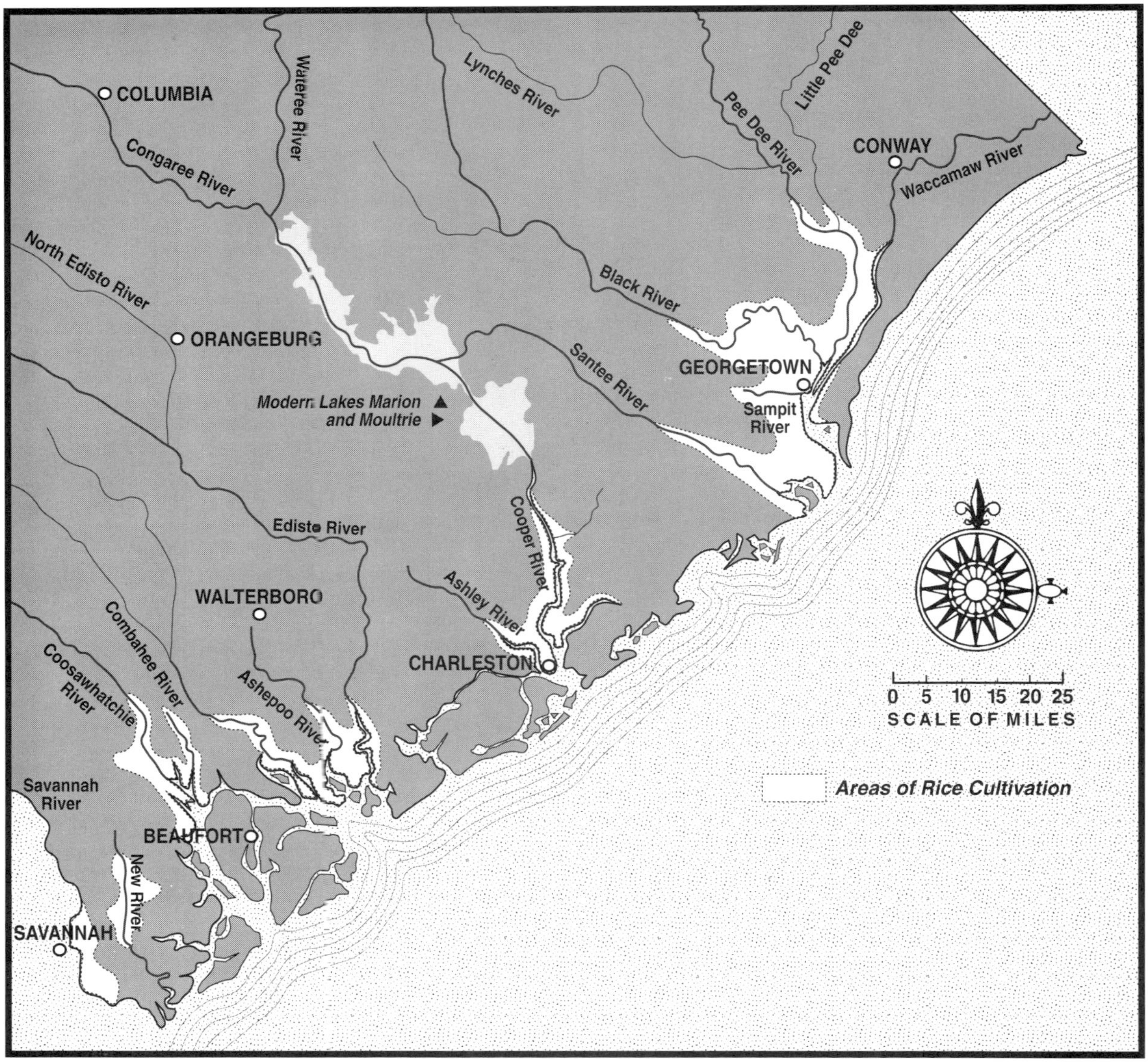

The South Carolina lowcountry showing areas of rice cultivation.

FROM THE WEST INDIES

Describing the South Carolina gentry in 1773 the *South Carolina Gazette* commented:

> Their whole Lives are one continued Race: in which everyone is endeavoring to distance all behind him; and to overtake or pass by, all before him. . . . Every Tradesman is a Merchant, every Merchant is a Gentleman, and every Gentleman one of the Noblesse. We are a Country of Gentry. . . . We have no such thing as a common People among us: Between Vanity and Fashion the species is utterly destroyed.[1]

Professor Richard Dunn uses this statement to suggest the character of a society whose acquisitive nature was established by conditioning in the West Indies. "Carolina planters of the eighteenth century," he avers, "had more in common with Barbados sugar planters of the seventeenth century than large gangs of slaves."

But many of the early settlers who left Barbados for the Carolina region did so because of an inability to compete with sugar planters and their large gangs of slaves. One wonders, therefore, how or why they, their descendants, or others who came after them created a society that so closely paralleled the one they had fled?

Top: Kinloch Plantation rice barn in Georgetown, South Carolina. Opposite page: Kinloch Plantation rice mill in Georgetown, South Carolina. (South Carolina Department of Archives and History, State Historic Preservation Office files.)

The answer, in a word, and somewhat too simply, is rice. For rice was the source of the wealth, the large plantation units, and the sizable slave holdings. Its cultivation created a world that differed from that of the Middle Atlantic tobacco growers. This world of great wealth required large plantation units and sizable slave holdings. The wealth from rice produced a one-crop agriculture that helped make coastal South Carolina distinctive and caused it to look more like Barbados or some other West India island. It also exacerbated the distance between rich and poor.

By the middle of the eighteenth century, according to historian Charles Joyner, Carolina rice planters enjoyed the highest per capita income in North America and perhaps the fastest per capita growth rate in the world.[2] By 1860, twenty-nine of the eighty-eight planters in the United States who owned over three hundred slaves produced rice; nine of the fourteen planters who owned more than five hundred slaves grew rice; and

Curing rice in South Carolina (South Carolina Department of Archives and History, photograph collection.)

the only person in the United States who owned more than one thousand slaves was a South Carolina rice planter—Joshua Ward of Waccamaw. Ward served in the South Carolina House of Representatives, in the Senate, and as lieutenant governor. His 1,092 slaves produced 3,900,000 pounds of rice in 1850. Ward developed a special big grain rice and received many honors for his high rice production.[3]

THE IMPORTATION OF SLAVES

Now what did this slave population mean practically? Well, for the rice growing regions it meant a demographic disproportion of blacks. "Only the African race," exclaimed the planter's daughter Elizabeth Allston Pringle, "could have made it possible or profitable to clear the dense cypress swamps and cultivate them in rice by a system of flooding the fields from the river by canals, ditches, or floodgates, drawing off the water when necessary, and leaving these wonderfully rich rice lands dry for cultivation."[4]

A twentieth century planter, David Doar, was less generous in attribution and placed responsibility for the development of rice culture more clearly with those of his own class. "As one views this vast hydraulic work [the rice plantation], he is amazed to learn that all of this was accomplished in face of seemingly insuperable difficulties by every-day planters who had as tools only the axe, the spade, and the hoe, in the hands of intractable negro men and women, but lately brought from the jungles of Africa."[5]

Despite the difference in emphasis, each writer calls attention to the role of blacks. For unlike the Chesapeake where slavery developed gradually, the proprietors provided for the institution in the Carolina region with the colony's inception in 1663. And while the black population grew slowly in the Chesapeake, the reverse was true in Carolina. Blacks were imported in significant numbers from about the 1690s, but by

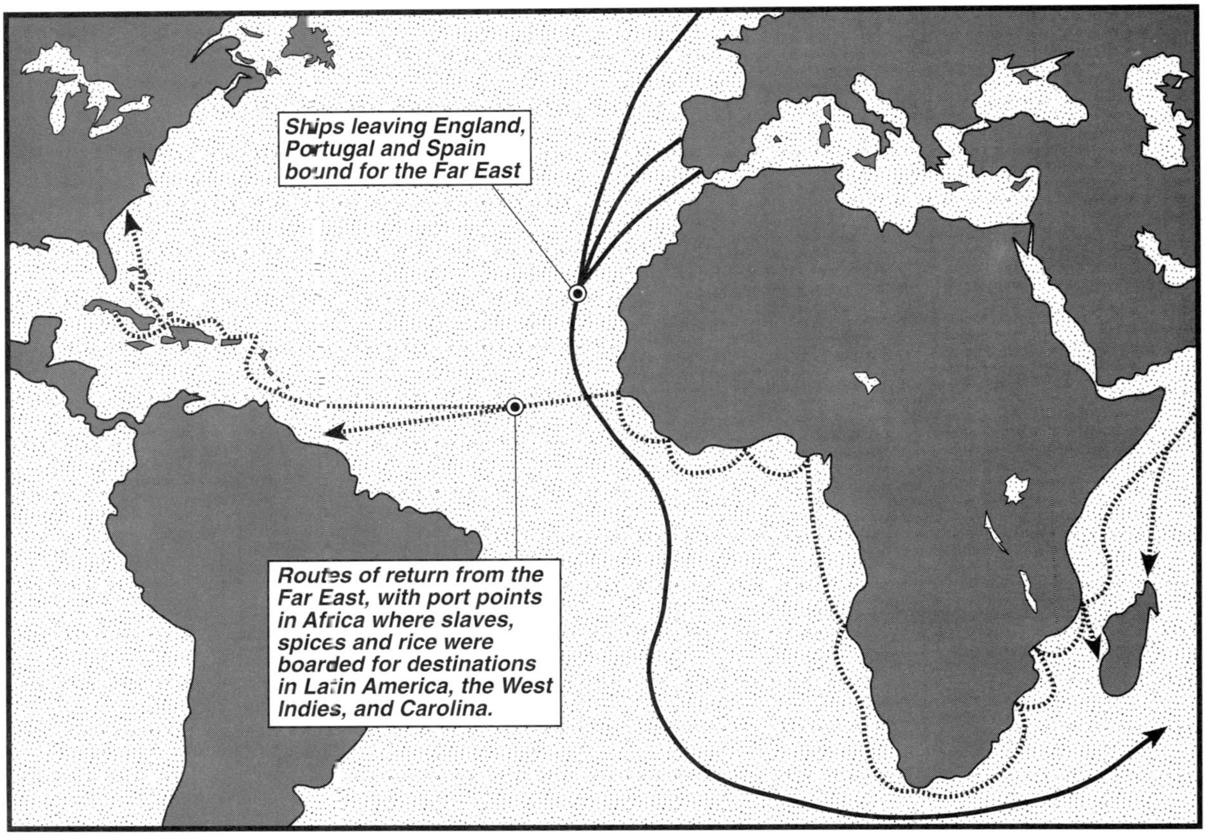

Ships left England, Portugal, and Spain bound for the Far East and returned from there to ports in Africa where slaves, spices, and rice were boarded for destinations in Latin America, the West Indies, and Carolina.

1715, the black population outnumbered the white by around 40 percent (10,500 to 6,250). South Carolina was the only colony in English North America where this disproportion existed. In the same years, the colony began the successful production of rice.[6]

It is worth considering whether, without rice cultivation, South Carolina might have developed in the direction of Virginia—containing a significant black population but with a white majority in the first half of the eighteenth century. This argument may not be entirely sustainable since differences would be expected, but it suggests that economic factors play a powerful role in determining patterns of development.

PIONEER CONDITIONS

As in Virginia, most seventeenth-century South Carolina slaves came from the West Indies, some accompanying the original English settlers from Barbados. Although the colder winters created some disadvantages for them, these blacks came physically and pharmacologically better equipped to cope with the new colony's semitropical environment than their masters . Their African background made them more disease resistant and provided useful cultural knowledge. They had a familiarity with tropical herbs, an ability to move along inland waterways using canoes or pirogues, and skill in fishing—using nets, harpoons, and other implements, including drugs or poisons, to stun fish. These capabilities, among others, enabled them to live off the land more easily than their masters could.

African expertise as well as the rough pioneer conditions of a new settlement made it easier for the slaves to obtain a degree of what has been termed "bucksaw" equality that did not mask their servile status but doubtless enhanced their self esteem—factors that may have created for blacks in Carolina conditions somewhat analogous to those existing for blacks in Virginia.

The northern colony's more temperate climate did not negate the usefulness of an African agricultural background. For while slavery was forming, Virginians made it possible for blacks to work their way to freedom and, thanks to prior skills, some accomplished the feat. They were certainly more capable than the

A Sea Island rice field. Taken from Harpers, *November 1878. (South Carolina Department of Archives and History, photograph collection.)*

English townsmen who found themselves suddenly in tobacco fields. Enslaved blacks often worked together with white indentured servants, and treatment of the two may not have been all that different.

In both colonies an inchoate society permitted rough equality among working blacks and whites.

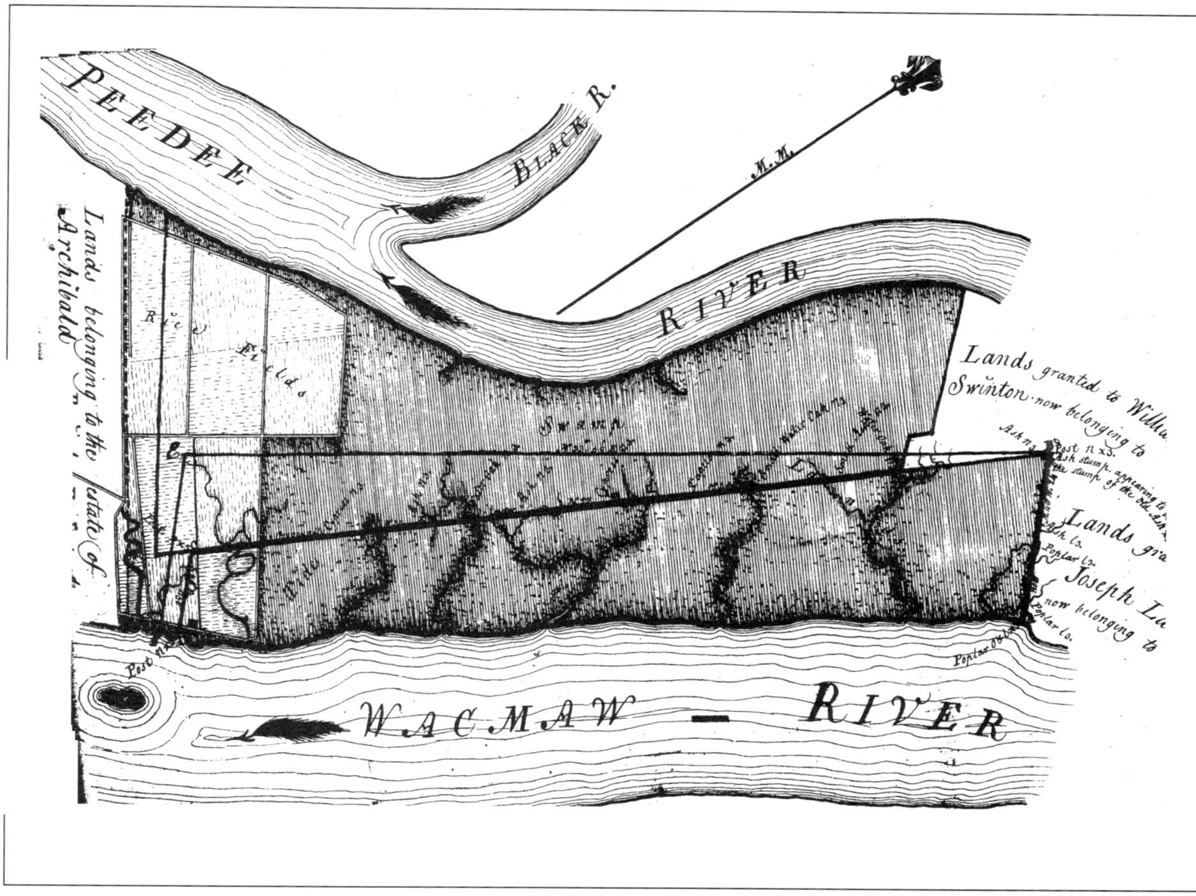

A 1787 survey of adjoining tracts of what became rice land situated between Peedee and Waccamaw rivers in Georgetown District, one belonging to William and Thomas Alston and the other to Francis and Cleland Kinloch. (South Carolina Department of Archives and History, Charleston District Court of Common Pleas, Judgment Rolls, 1788, # 223 A.)

Legislation enacted after Bacon's Rebellion in 1676 brought this situation to an end in Virginia. The South Carolina slave code of 1696, based on the Barbadian code of 1688, announced the end of this relatively benign period in the lowcountry. Both colonies increasingly embraced African labor and South Carolina increasingly embraced rice.[7] There the similarity ends.

PRODUCTIONS OF AGRICULTURE, ETC.—Continued.

STATES AND TERRITORIES.	Value of live stock.	Value of animals slaughtered.	PRODUCE DURING THE YEAR ENDING JUNE 1, 1850. Wheat, bushels of.	Rye, bushels of.	Indian corn, bushels of.	Oats, bushels of.	Rice, pounds of.	Tobacco, pounds of.	Ginned cotton, bales of 400 pounds each.
Maine	$9.705,726	$1,646.773	296,259	102,916	1,750.056	2,181.037	....	....	....
New Hampshire	8.871,901	1.522,873	185,658	183,117	1,573.670	973,381	....	50	....
Vermont	12.643,228	1.861,336	535,955	176.233	2.032,396	2,307,734	....	....	....
Massachusetts	9.647,710	2,500,924	31,211	481,021	2,345,490	1,165,146	....	138,246	....
Rhode Island	1.532,637	667,486	49	26,409	539,201	215.232	....	....	....
Connecticut	7.467,490	2,202,266	41,762	600,893	1.935.043	1.258,738	....	1,267,624	....
New York	73.570,499	13,573,883	13.121,498	4,148,182	17.858,400	26.552,814	....	83,189	....
New Jersey	10.679,291	2,638,552	1.601,190	1.255,578	8,759,704	3,378.063	....	310	....
Pennsylvania	41.500,053	8,219.848	15,367,691	4,805.160	19.835,214	21.538,156	....	912,651	....
Delaware	1.849,281	373,665	482.511	8.066	3,145,542	604,518	....	....	....
Maryland	7.997,634	1,954.800	4,494,680	226,014	10,749,858	2,242,151	....	21,407,497	....
District of Columbia	71.643	9,038	17.370	5,509	65,230	8,134	....	7,800	....
Virginia	33.656.659	7.502.986	11,212.616	458.930	35,254,319	10,179,144	17,154	56,803.227	3.947
North Carolina	17.717,647	5,767.866	2,130.102	229.563	27,941,051	4.052.078	5,465.868	11,984.786	73.845
South Carolina	15.060,015	3.502,637	1.066.277	43.790	16,271,454	2,322,155	159,930,613	74,285	300.901
Georgia	25.728,416	6,339,762	1,088.534	53,750	30.080.099	3,820.044	38,950.691	423.924	499.091
Florida	2.880,058	514,685	1.027	1,152	1.996.809	66,586	1,075.090	998.614	45.131
Alabama	21.690,112	4.823,485	294.044	17.261	28.754.048	2.965,696	2,312.252	164.990	564.429
Mississippi	19.403.662	3,636.582	137,990	9,606	22.446.552	1,503.288	2,719.856	49.960	484.292
Louisiana	11.152,275	1.458,990	417	475	10.266.373	89.637	4,425.349	26.878	178.737
Texas	10.412.927	1.116,137	41.729	3.108	6.028.876	199,017	88.203	66.897	58.072
Arkansas	6.647,969	1,163,313	199.639	8.047	8.893.939	656,183	63.179	218.936	65.344
Tennessee	29.978,016	6,401,765	1,619,386	89.137	52,276,223	7,703,086	258,854	20,148.932	194.53
Kentucky	29.661,436	6.462,598	2,142.822	415.073	58.672,591	8,201,311	5.688	55,501.196	75
Missouri	19.887,580	3,367,106	2,981,652	44.268	36,214.537	5,278.079	700	17,113,784	....
Illinois	24.209,258	4,972,286	9,414.575	83.364	57,646,984	10,087,241	....	841.394	....
Indiana	22.478,555	6,567,935	6.214,458	78.792	52.964.363	5,655,014	....	1,044.620	1
Ohio	44.121,741	7,439,243	14.487.351	425.918	59.078,695	13,472,742	....	10,454.449	....
Michigan	8.008,734	1,328,327	4.925.889	105.871	5.641,420	2.866.056	....	1.245	....
Wisconsin	4.897.385	920,178	4.286,131	81.253	1.988.979	3,414.672	....	1.268	....
Iowa	3.689,275	821.164	1,530.581	19.916	8,656,799	1,524,345	....	6.041	....
California	3.351.058	107,173	17,328	....	12.236	....	....	1,000	....
Minnesota Territory	92.859	2.840	1.401	125	16,725	30,582	....	....	....
New Mexico Territory	1.494.629	82,125	196.516	....	365,411	5	....	8,467	....
Oregon Territory	1.876,189	164,530	211.943	106	2.918	61,214	....	325	....
Utah Territory	546.968	67,985	107,702	210	9.899	10,900	....	70	....
Total	544.180,516	111,703,142	100,485.944	14,188,813	592,071,104	146,584,179	215.313,497	199,752,655	2,469,09

*A portion of one page of the Agricultural Census for 1850 shows that South Carolina was by far the largest producer of rice. (*The Seventh Census of the United States: 1850, *lxxii.)*

ETHNICITY

Lowcountry planters benefited from the knowledge of their African bondsmen, many of whom knew more about rice cultivation than did their owners. Englishmen had no background in producing the crop and had difficulty making it grow. Africans had the expertise that Englishmen lacked, and the suggestion is strong that they taught Englishmen how to raise the crop. Elizabeth Pringle may have said more than she knew when she linked Africans and the development of rice culture. Indeed, Carolinians adopted a preference for peoples from the rice-producing Senegambia region that lasted through most of the colonial period, although the vagaries of trade prevented that region's ethnic groups from always dominating importation statistics. In 1765, blacks outnumbered whites by more than two to one (90,000 to 40,000), and Charleston was the largest North American slave port. By that date, South Carolina looked more like Barbados than it had ever looked like Virginia.

The force and significance of this demographic fact and some of its picturesque implications are suggested by advertisements for slave runaways. "Lately run away from my plantation at Ashepoo," James Parsons wrote in the *South Carolina Gazette*, 24 October 1761:

> two tall likely young new *Bambara* negro fellows, named *Abram* and *Lymat*. . . . Also run away . . . about two years ago, a short chubby hairy *Angola* fellow named *March*, and about two months

LATELY run away from my plantation at *Aſhepoo*, two tall likely young new *Bambara* negro fellows, named *Abram* and *Lymat*, each of them had on a crocus frock without ſleeves, a pair of black ſtocking breeches, and oſnaburgh trowſers. Alſo run away from the ſaid plantation, about two years ago, a ſhort chubby hairy *Angola* fellow named *March*; and about two months ſince a very black ſlip ſhod fellow, this country born, called *Harry*, who ſpeaks good *Engliſh*, and is well known in *Charles-Town*, and many places in the Country: Likewiſe went away about this time twelvemonth (two days after I bought him in the ſhip in which he was imported) from Mr *William Williamſon*'s plantation in St. *Paul*'s pariſh a new negro lad of the *Pappa* country, his country name *Arrow* & will pay *Ten Pounds* reward, and all legal charges, for each of the ſaid negroes that ſhall be takm up and delivered to my overſeer at my ſaid plantation, or to me in *Charles Town*; and *Two Hundred Pounds* reward to any perſon that ſhall inform againſt a white man, and *Twenty Pounds* to any informing againſt a ſlave, harbouring any one of the ſaid negroes upon conviction of the harbourer.

JAMES PARSONS.

James Parsons' advertisement for runaway slaves in the 24 October 1761 South Carolina Gazette.

Hoeing rice. Harpers, *November 1878. (South Carolina Department of Archives and History, photograph collection.)*

since a very black slip shod fellow, this country born, called *Harry*. . . . Likewise went away about this time twelve month . . . a new negro lad of the Pappa country named *Arrow*. . . . [8]

Virginians also identified runaways in ethnic terms when necessary but not to the extent that was common in South Carolina. Neither blacks in general nor the African-born in particular were as great a proportion of the population there. Nowhere else among the English in North America was this scene replicated on a comparable scale.

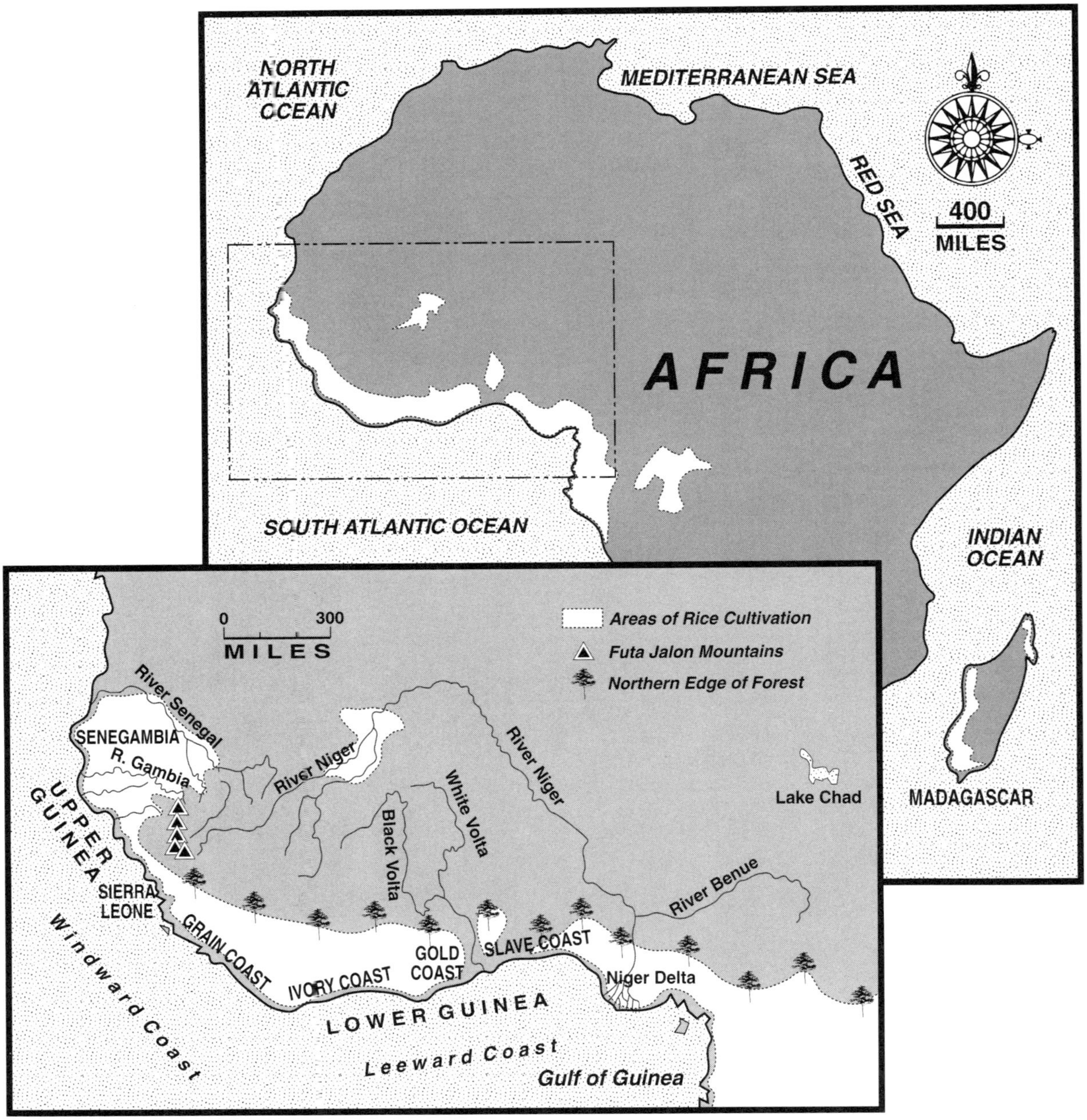

NORTH ATLANTIC OCEAN
MEDITERRANEAN SEA
RED SEA
400 MILES
AFRICA
SOUTH ATLANTIC OCEAN
INDIAN OCEAN
MADAGASCAR
0 300 MILES
Areas of Rice Cultivation
Futa Jalon Mountains
Northern Edge of Forest
River Senegal
SENEGAMBIA
R. Gambia
River Niger
Black Volta
White Volta
River Niger
River Benue
Lake Chad
UPPER GUINEA
SIERRA LEONE
GRAIN COAST
IVORY COAST
GOLD COAST
SLAVE COAST
Niger Delta
LOWER GUINEA
Windward Coast
Leeward Coast
Gulf of Guinea

These surroundings could not help but affect the cognitive processes of white South Carolinians, and the region was and is unique in maintaining an awareness of African ethnic differences. Like Virginians, Carolina planters ultimately preferred to work with native-born laborers, who, as Governor John Glen once expressed it, "may on all occasions be more relyed on than Guinea slaves."[9] But where African-born slaves were concerned, they maintained their regional biases into the nineteenth century.

Thus Charleston merchant Christopher Fitzsimmons, trading in the West Indies in 1807, not many months before the termination of the Atlantic slave trade, instructed his agent to bring back new Africans. "In the purchasing of slaves," he ordered, "you will be careful in the choice of them as well as in appearance and health as also to be of [an] approved nation." He also indicated that he could tell the difference. "I went on board with a friend to examine the cargo," he reported, "and found the negroes looked healthy and in good order but very diminutive and not of an approved nation for the planters of this country."[10]

The slaves made their own ethnic evaluations. A nineteenth century bondsmen did so this way: "Peter and Sampson and David, dem ben an outlan' people Afrikan, one ben Gullah and one ben a Guinea—the Gullah ben a cruel people—and de Fullah be a cruel people, but Guinea ben a rough workin' people, and Milly ben a Guinea."[11]

This awareness was reflected in the twentieth century by a white Carolinian who tried to discuss race relations in terms of the mixed character of the black population. In this case he meant a mixture of African ethnic groups rather than a mixture between blacks and whites. "The negro population of the United States," he said, "is probably as much mixed as the white population."[12]

Opposite page: Coastal Africa showing the areas of rice cultivation. Carolinians adopted a preference that lasted through most of the colonial period for peoples from the rice-producing Senegambia region.

CULTURAL PRESERVATION AND TRANSFORMATION

While rice was less labor-intensive than sugar, the scale of labor required made it more labor intensive than tobacco. By limiting production to the wealthy, it established an aristocratic cast to South Carolina society. To begin a rice plantation under the most favorable conditions, planters needed about thirty slaves. Moreover, because these slaves worked in larger production units than was characteristic of the Chesapeake, they kept in closer contact with other blacks than they did with the master or overseer.

These conditions helped the slaves preserve a measure of African culture at the same time they engaged in cultural transformation. And the rapid importation of large numbers of new Africans—the so-called "salt-water" slaves—within a relatively short period of time helped to preserve the African nature of slave culture.

Despite attempts of the South Carolina legislature to mandate a set ratio of white to black people, whites were in short supply on some plantations. Consequently, blacks were given more responsibility. A Georgia merchant described two families purchased in Carolina as "hav[ing] been born on . . . [the owner's] plantations, [and] are likely young people, well acquainted with Rice & every kind of plantation business, and in short [are] capable of the Management of a plantation themselves."[13]

Contemporary sources suggest that black foremen, or drivers, were unique to the South Carolina rice country during the colonial period—though they did foreshadow the basic organizational pattern of the large antebellum plantation. This circumstance of blacks spending much time among themselves, even when they were working for whites, explains why one historian has described colonial South Carolina as being essentially bi-lingual, with blacks speaking a patois that most whites did not understand.[14] This creole language survived into the twentieth century as the Gullah or Geechee still spoken along the South Carolina-Georgia coastline.

Three scenes of Mansfield Plantation: Slave chapel (top); typical slave house [ca.1830–ca.1860] (below left); and oak avenue with an intact street of slave cabins (below left). (South Carolina Department of Archives and History, State Historic Preservation Office files.)

CROSS CULTURATION

This developing African American culture influenced the whites as whites influenced blacks. Thus, Charles Joyner relates, "Growing up on her father's Georgetown rice plantation, Elizabeth Allston Pringle eagerly absorbed the stories 'Daddy Tom', and 'Daddy Prince' and 'Maum Maria' told her of their own childhoods in Africa. It was, she said, a 'very peculiar life, surrounded by hundreds of a different race.'"[15]

Interchange between blacks and whites occurred in numerous ways and was quite pronounced in language. In this instance South Carolina was not unique. Eighteenth- and nineteenth-century observers made frequent comment on the fact. The English visitor Sir Charles Lyell, travelling in up-country Georgia in 1846, met a family in which black and white children had been brought up together and where the white children—illegally—had taught the black children how to read. "Unfortunately," he remarked, "the whites, in return, often learn from the negroes to speak broken English, and, in spite of losing much time in unlearning ungrammatical phrases, well-educated people retain some of them all their lives."[16] A previous English visitor had made much the same comment a century earlier. English actress Fanny Kemble thought that what she called the "Negro mode of talking" was "distinctly perceptible in the utterances of all Southerners," and she was distressed to find the process in operation with her own daughter:

> "I am amused," she said, "but by no means pleased at an entirely new mode of pronouncing which S[ally] has adopted. Apparently the Negro jargon has commended itself as euphonious to her infantile ears, and she is now treating me to the most ludicrous and accurate imitations of it every time she opens her mouth. Of course I shall not allow this, comical as it is, to become a habit. This is the way the Southern ladies acquire the thick and inelegant pronunciation which distinguishes their utterances from the northern snuffle, and I have no desire that S[ally] should adorn her mother tongue with either peculiarity.[17]

But in a situation where blacks predominated and whites were surrounded by them, the process was unavoidable. Professor George C. Rogers, perhaps with tongue in cheek, has referred to the modern Charleston accent as "high Gullah."[18]

Two views of Chicora Wood. The main house, built by 1820, was originally called Matanzas, and was owned by prominent rice planter (and governor) Robert F. W. Allston, whose daughter, Elizabeth Allston Pringle, lived a "very peculiar life, surrounded by hundreds of a different race." (South Carolina Department of Archives and History, State Historic Preservation files.)

THE TASK SYSTEM

Rice was produced using the "task system." Rather than working gangs from dawn to dusk in assembly-line fashion characteristic of labor associated with sugar or cotton, the task system permitted a slave to complete an assigned amount of labor, varying from one quarter to a half acre, depending on the job, and then left him free to work for himself.

THE DEVELOPMENT OF THE TASK SYSTEM

How the task system developed in Carolina is not entirely clear. West Indian settlers who came to the mainland left a region where gang labor prevailed. Even in the islands, however, there were some jobs that seemed suited to tasking.

As a general rule, one scholar suggests, tasking could be used with hardy crops that did not require minute supervision—like rice; or in cases where a production schedule was not urgent; or where the labor force was scattered, making the close supervision that demanded gang labor difficult. In the case of rice, he seems to imply, the character of the crop dictated the system, but he does not indicate whether master or slave first perceived the need.[19]

Another scholar places the initiative with the slave. The task system arose, he suggests, because wealthy rice planters inclined towards absenteeism, and their lack of minute supervision permitted slaves, influenced by their African background, to establish a practice in which they saw some advantage.[20]

Still a third historian blames the environment. She writes that "constant exposure to heat, humidity, dampness, insects and poisonous reptiles made cultivating rice, without doubt, more objectionable and strenuous than cultivating dry culture crops and must account for the practice of tasking for work assignments."[21]

There is something to be said for all these explanations, and they all may be, in some measure, contributory. Whatever the precise lines of evolution, planters doubtless saw tasking as an advantage. It also benefited the slave and had important consequences for slave culture. The procedure became highly identified with, though not exclusive to, the South Carolina-Georgia lowcountry; and it was used with sea-island (or long-staple) cotton as well as with rice.

Bcause it provided slaves with significant amounts of time for themselves, tasking gave them a measure of independence—they could work their own garden plots, hunt, fish, or do other jobs that contributed to their personal welfare. One former field hand commented: "a good active industrious man would finish his task sometimes at 12, sometimes at 1 and 2 oclock and the rest of the time was his own to use as he pleased." Another said "I could save for myself sometimes a whole day[;] if I could do 2 tasks in a day then I had the next day to myself. Some kind of work I could do 3 tasks in a day."[22]

TASKING AND MARKETS

Tasking also permitted slaves the possibility of accumulation, for it fueled an internal market system not unlike that in West India islands. In both Jamaica and South Carolina, for example, slaves traded in the marketplace, and in both areas masters purchased slave produce.

The South Carolina Grand Jury objected in 1734 to what it called the "intollerable Hardship" brought on by the fact that "Negroes are suffered to buy and sell and be Hucksters of Corn, Pease, Fowls etc. whereby they watch Night and Day on the several Wharves, and buy up many Articles necessary for the Support of the Inhabitants, and make them pay an exorbitant Price for the same. . . . " The General Assembly attempted to outlaw the practice in 1730 but without effect. The practice was still being publicly protested in 1742, and the runaway slave Bella was described as having been "almost every Day in the Market selling divers Things." At the end of the century, the legislature admitted defeat. It abandoned the attempt to prohibit slave marketing, conceded slaves the right to buy and sell their own goods, and sought merely to regulate the practice.

TASKING AND COMMUNAL ACTIVITY

While tasking spurred individual initiative, it also permitted communal activity, for the work songs that signaled gangs of slaves moving in unison could also be used in tasking. Singing represented maintenance of an African communal attitude towards labor and helped slaves, particularly slower ones, keep pace, even though they completed tasks individually.

The system also enhanced family unity, for husband could help wife, or vice versa, if either finished a task first. Moreover, children might help parents. In the best of conditions, it reenforced the family as an economic unit because for at least part of the working day, the family could work together and for themselves.

And slaves who demonstrated a talent for management and gained the respect of the slave community by their steadiness and efficiency often assumed leadership roles in freedom.

PROS AND CONS OF TASKING

The tasking system benefited planters, for it encouraged slaves to complete an assignment quickly and well and attached them to the plantation as well. As one proclaimed, "no Negro with a well-stocked poultry house, a small crop advancing, a canoe partly finished or a few tubs unsold, all of which he calculates soon to enjoy, will ever run away."[23] It also benefited slaves, for they were able to accumulate money and property and to validate their sense of self-worth.

Nevertheless, slavery was a constant tug-of-war between slave and master. Some slaves complained that masters did not honor the task arrangement properly because they routinely created tasks that most slaves could not complete early; and some planters roundly resented the independence the system provided and tried to constrain it in various ways. But the consequences of the constraints were so disruptive of plantation harmony in general and of a productive work regimen in particular that few planters attempted to alter the accepted custom for long.[24]

The concessions that lowcountry (and other) planters made to the slave's ability to determine his own work pace, to choose his own avocations, to accumulate his own property, and to engage in his own marketing, recognized the slave's humanity. On the other hand, the extent to which a slave became satisfied with his situation—the object at which the planter aimed—made him an unwitting conspirator in his own enslavement.

A couple ploughing rice. Harpers, *November 1878. (South Carolina Department of Archives and History photograph collection.)*

INTERRACIAL MINGLING

Rice wealth permitted the absenteeism that an unhealthy environment encouraged. It also allowed a sumptuous lifestyle and a hint of dissipation. In the English West Indies where blacks outnumbered whites in a range of from three to ten to one, slave concubinage became a custom. There, a planter was almost expected to take a black or mulatto mistress. This practice eventuated in a caste system based on color, where lighter-skinned people gained rights and privileges denied to blacks. Though a similar gradation was never institutionalized among the English in North America, nor did the taking of slave mistresses achieve such open acceptability, they came closer to doing so in South Carolina than anywhere else.

A critic complained in 1772 that:

> to such a Pitch of Licentiousness have some Men arrived, that they cohabit, as Husband, with Negro Women, treating them as Wives even in public, and do not blush to own the Mongrel Breed...which is thus begotten; maintaining the spurious Progeny, as well as the Mother, in Splendor, whilst many virtuous and industrious white Women might be found, who, though in low Circumstances, would be a Credit to a Man, and much less expensive. . . .

He thought that "this *scandalous* Intimacy, which too much subsists between the Sexes of different

Colours," was "a principal Reason why the Wenches are so insolent and useless to their Owners, as well as the Cause of their frequent Elopements, during which they are concealed and supported by their abject Paramours."

And Josiah Quincey, Jr., a New Englander who travelled through South Carolina in 1773, also disapproved of "the intercourse between the whites and blacks. The enjoyment of a negro or mulatto woman," he wrote,

> is spoken of as quite a common thing: no reluctance, delicacy or shame is made about the matter. It is far from being uncommon to see a gentleman at dinner, and his reputed offspring a slave to the master of the table. I myself saw two instances of this, and the company very facetiously would trace the lines, lineaments and features of the father and mother in the child, and very accurately point out the more characteristic resemblance. The fathers neither of them blushed or seem[ed] disconcerted. [25]

This interracial dalliance was usually between white men and black women; seldom between black men and white women, though that kind of relationship was by no means unknown. White women occasionally protested the prevailing practice, however, as Charles Myers discovered. He was forced publicly to "give notice afresh, to all persons whatsoever, not to harbour, entertain, or credit his Wife, *Mary Myers,* who some time ago eloped from him; and notice thereof was given in the public Gazette, forwarning all persons from harbouring, entertaining, or giving her credit in his name, as he is determined not to pay any debts of her contracting; and to prosecute, to the utmost rigour of the law, all such who shall harbour or entertain her."

This notice was little different in tone from those posted for runaway slaves or indentured servants except that in those cases a reward was commonly offered for the fugitive's return. The papers seldom had occasion to print the fugitive's account directly. But Mary Myers, clearly a feisty individual, determined to present her story to a candid world. A few weeks later she responded: "A MAN who forces his Wife from him, by inhumanity, and preferring an old NEGRO WENCH for a BEDFELLOW, certainly cannot deserve credit. CHARLES MYERS, who advertises against crediting or entertaining his Wife might therefore have spared himself that trouble and expence." [26]

Unlike the situation in most southern states and in harmony with that in many of the islands, interracial marriages were not illegal in antebellum South Carolina. This may have been because, as one historian suggests, few people in the islands would have considered such a thing.[27] Or it may have been because too many people knew or suspected too much.

John Ross, plantation manager, writing from St. Augustine where his loyalist sentiments confined him at the outbreak of the American Revolution, may have expressed the attitude prevalent in both regions.

Many planters built Charleston-style summer homes in Pendleton to take advantage of open space and summer breezes. Woodburn, with its large rooms, extremely high ceilings, and central hallways, is typical. (South Carolina Department of Archives and History, State Historic Preservation Office files.)

Concerning a black woman with whom he had taken up, he assured his father that "I wonder you should suspect me of any other connexion with such a Wench, than that of having got some children by her. I am not yet Old enough for dotage, altho my head & beard are become pretty gray." He did, however, at considerable cost, purchase his mistress from her owner and freed and sent his children by her to Scotland.[28]

Thus, there developed in nineteenth-century Charleston, as in Kingston or New Orleans, a class of free colored people allied in sentiment and by consanguinity to the master class. And though some of these people were immigrants from St. Domingue who fled the revolution against the French, the society into which they came had a place for them already.

THE CULTURAL HYBRID

Rice contributed to the making of Carolina as a cultural hybrid. Massive white immigration into the interior in the second half of the eighteenth century shifted the balance of power towards the interior and obliged planter oligarchs to consider the attitudes and outlooks of farmers and laborers bred on the continent. But rice, its wealth, and the ramifications of its labor needs insured that the imprint of the islands would not be effaced. ■

NOTES

1. Richard S. Dunn, "The English Sugar Islands and the Founding of South Carolina," *South Carolina Historical Magazine* 72 (April 1971), 93.

2. Michael Mullin, ed., *American Negro Slavery: A Documentary History* (New York: Harper and Row, 1976), 10.

3. Charles Joyner, *Down By The Riverside: A South Carolina Slave Community* (Urbana: University of Illinois Press, 1984), 34. Louise Bailey, et al., eds., *Biographical Directory of the South Carolina Senate, 1776–1985* (Columbia: University of South Carolina Press, 1986), III: 1674–5.

4. Elizabeth W. Allston Pringle, *Chronicles of Chicora Wood.* (New York: Scribner's Sons, 1922), 15.

5. David Doar, *Rice and Rice Planting in the South Carolina Low Country.*, ed. E. Milby Burton (Charleston, S.C.: The Charleston Museum, 1936) as quoted in Karen Hess, *The Carolina Rice Kitchen: The African Connection* (Columbia: University of South Carolina Press, 1922), 9.

6. The best single source on the development of rice cultivation in South Carolina is Peter Wood *Black Majority: Negroes in Colonial South Carolina from 1670 through the Stono Rebellion* (New York: Alfred A Knopf, 1974), esp. 35–62. Also see Daniel C. Littlefield, *Rice and Slaves: Ethnicity and the Slave Trade in*

Colonial South Carolina (Baton Rouge: Louisiana State University Press, 1981; reprinted Urbana: University of Illinois Press, 1991), 74–114.

7. The inchoate nature of slavery in early Virginia can be viewed most succinctly in Timothy Breen and Stephen Innes, *"Myne Owne Ground": Race and Freedom on Virginia's Eastern Shore, 1640–1676* (New York: Oxford University Press, 1980). But also see Edmund Morgan, *American Slavery, American Freedom: The Ordeal of Colonial Virginia* (New York: W.W. Norton and Company, 1975); Timothy Breen, "A Changing Labor Force and Race Relations in Virginia, 1660–1710" *Journal of Social History* 7 (1973), 3–25; and Russell R. Menard, "From Servants to Slaves: The Transformation of the Chesapeake Labor System," *Southern Studies* 16 (1977), 355–90. The image of "bucksaw equality" is suggested by an incident in Peter Wood's *Black Majority*. A convenient little book which summarizes much of the literature on blacks in early British North America is Donald R. Wright, *African Americans in the Colonial Era: From African Origins Through the American Revolution* (New York: Harlan Davidson, 1990). Also see Leland Ferguson, *Uncommon Ground: The Archaeology of Early African America, 1650–1800* (Washington, D.C.: Smithsonian Institution Press, 1992).

8. Littlefield, *Rice and Slaves*, 115. There is an extended discussion of the significance of this demographic fact 115–73. See also Lathan A. Windley, comp., *Runaway Slave Advertisements: A Documentary History from the 1730s to 1790*, v. 3, *South Carolina* (Westport, Conn: Greenwood Press, 1983).

9. Daniel C. Littlefield, "'Abundance of Negroes of that Nation': The Significance of African Ethnicity in Colonial South Carolina," in David R. Chesnutt and Clyde N. Wilson, eds., *The Meaning of South Carolina History: Essays in Honor of George C. Rogers, Jr.* (Columbia: University of South Carolina Press, 1991), 19–38.

10. *Ibid.*, 22.

11. Joyner, *Down By The Riverside*, 39.

12. Littlefield, *Rice and Slaves*, 1.

13. *Ibid.*, 141.

14. Mullin, ed., *American Negro Slavery*, 10–17 and Daniel C. Littlefield, "Continuity and Change in Slave Culture: South Carolina and the West Indies," *Southern Studies* 26 (Fall 1987), 202.

15. Joyner, *Down By The Riverside*, 15.

16. Charles Lyell, *A Second Visit to the United States of North America*, (New York: Harper & brothers, 1850) II, 20.

17. Frances Anne Kemble, *Journal of a Residence on a Georgia Plantation in 1838–1839* (Athens: University of Georgia Press, 1984), 280–81.

18. Robert M. Weir, *Colonial South Carolina—A History* (Millwood: KTO Press, 1983), 198.

19. Philip D. Morgan, "Task and Gang Systems: The Organization of Labor on New World Plantations," in Stephen Innes, ed., *Work and Labor in Early America* (Chapel Hill: University of North Carolina Press, 1988), 189–220.

20. Ira Berlin, "Time, Space, and the Evolution of AfroAmerican Society in British Mainland North America," *American Historical Review* 8 (February 1980), 44–78.

21. Julia Floyd Smith, *Slavery and Rice Culture in Low Country Georgia, 1750–1860* (Knoxville: University of Tennessee Press, 1985), 50.

22. See Littlefield, "Continuity and Change in Slave Culture, 202–16; quotations on pp. 203–4.

23. Philip D. Morgan, "Work and Culture: The Task System and the World of Lowcountry Blacks, 1770 to 1880," *William and Mary Quarterly*, 3rd Ser., 39 (October 1982), 586, 597.

24. See Drew Gilpin Faust, *James Henry Hammond and the Old South: A Design for Mastery* (Baton Rouge: Louisiana State University Press, 1982), 69–104, for an example of a planter who tried to abolish the task system, ultimately without success.

25. Josiah Quincy, Jr., "The Journal of Josiah Qunicy, Junior, 1773," *Massachusetts Historical Society Proceedings*, XLIX (June 1916), 463.

26. See *South Carolina Gazette*, October 31, 1768 and November 21, 1768.

27. Michael P. Johnson and James L. Roark, *Black Masters: A Free Family of Color in the Old South* (New York: W.W. Norton and Company, 1984), 53, 169. Winthrop D. Jordan, *White Over Black: American Attitudes Toward the Negro, 1550–1812* (Chapel Hill: University of North Carolina Press, 1968), 140. Also see his "American Chiaroscuro: The Status and Definition of Mulattoes in the British Colonies," *William and Mary Quarterly*, 3rd. Ser., 19 (1962), 183–200.

28. Littlefield, "Continuity and Change in Slave Culture," 215.

OTHER READINGS ON RICE

Chaplin, Joyce E. "Agricultural innovations & modernity in the lower South, 1730–1815." In *An Anxious Pursiut.* Williamsburg: UNC Press for Institute of Early American History & Culture, ca.1993.

Coclanis, Peter A. "Economic life & death in the South Carolina low country, 1670–1920." In *The Shadow of a Dream.* New York: Oxford University Press, 1989.

Dethloff, Henry C. *A History of the American Rice Industry, 1685–1985.* College Station: Texas A & M Press, 1988.

Easterby, J. H. ed. *The South Carolina Rice Plantation as Revealed in the Papers of Robert F. W. Allston.* Chicago: University of Chicago Press, 1945.

Hess, Karen. *The Carolina Rice Kitchen: The African Connection* (Columbia: University of South Carolina Press, 1992).

Heyward, Duncan Clinch. *Seed from Madagascar.* Columbia: University of South Carolina Press, 1993.

Lachicotte, Alberta Morel. *Georgetown Rice Plantations.* Georgetown, S.C.: Georgetown County Historical Society, 1993.

Lawson, Dennis T. *No Heir to Take Its Place: The Story of Rice in Georgetown County, South Carolina.* Georgetown, S.C.: The Rice Museum, 1972.

Pringle, Elizabeth W. Allston. *A Woman Rice Planter.* Columbia: University of South Carolina Press, 1992.

Rogers, George C., Jr. *The History of Georgetown County, South Carolina.* Columbia, S.C.: University of South Carolina Press, 1970.

Salley, A.S. *The Introduction of Rice Culture into South Carolina.* Columbia, S.C.: State Co., 1919.

Smith, Alice R. Huger. *A Carolina Rice Plantation of the Fifties.* New York: W. Morrow and Company, 1936.

Vernon, Amelia Wallace. *African Americans at Mars Bluff, South Carolina.* Baton Rouge: Louisiana State University Press, 1993.